Day Trading Strategies:

20 Golden Lessons to Start Trading Like a PRO Today! Learn Stock Trading and Investing for Complete Beginners. Day Trading for Beginners, Forex Trading, Options Trading & more

2

Table of Contents

Introduction

Congratulations on purchasing *Day Trading Strategies: 20 Golden Lessons to Start Trading Like a PRO Today!* and thank you for doing so. The world of trading is growing increasingly chaotic. Downloading this book is the first step that you can take towards doing something about your financial situation. The first step will not always be the easiest, which is why the information you will find in the following chapters is so important to take to heart, as they are not concepts that can be put into action immediately. If you file these concepts away for when you need them, when the time comes to actually use them, you will be glad you have them at hand.

The following chapters will discuss the primary preparedness principles that you will need to consider if you ever hope to really make money on day trading. This means that you will want to consider the quality of your entry and stop loss, including the potential issues raised by their ratio, how they can best be utilized in a strategy, as well as various tools you might need to keep your mind focused on the task at hand.

With those out of the way, you will then learn everything you need to know about money management. Rounding out the three primary requirements for successful day trading, you will then learn about crucial techniques that will help you in your journey.

I am happy to welcome you to the world of day trading and to help you make more money.

Chapter 1: Getting Started

In this book, we will show you all the steps you need to take to invest in forex from home. We will show you how to play on the forex market, how to choose the best pairs to invest in, and above all, how to invest in forex and currencies.

In addition, we will also mention the possible methods to invest in online foreign exchanges thanks to online trading. All the concepts that you will find in this guide have been written to be understood also by people unrelated to the world of online trading and the stock exchange, who has decided to inquire to start investing in the stock market.

Investing in the forex market means buying and selling currencies, aiming to earn between the price difference (purchase and sale). In the world of the forex exchange, as in the other major financial markets (for example, the stock market and CFD), you can earn both when there is an increase in the value of a stock and when there is a fall in the value of a stock.

Today, thanks to online trading, it is possible

to invest in forex simply from home without problems. This is possible thanks to online trading platforms better known as brokers.

Today, it is possible to invest in the foreign exchange mainly through the following methods:

1.   forex market;
2.   binary options trading;
3.   CFD trading (contract for difference)

In this case, you can choose one of the following online, regulated, and authorized trading platforms as shown below:

1.   Markets.com
2.   24Option.com
3.   iqoption.com
4.   BDSwiss.com

In short, with online trading, everyone can start making money on the forex market. It does not matter whether you are a novice trader or an experienced trader. Online trading is offered to everyone, thanks to the training offered by its broker, which teaches the basics of trading. Moreover, many brokers today allow you to practice with a demo account, which allows all traders to test not only on the trading platform, but they can also start experimenting with their trading strategies and take their first steps in this fantastic world.

Chapter 2: Trading or Saving?

Very often, the concepts of saving and investing are often confused, as well as that of "saver" and "trader". However, there are substantial differences that need to be understood before diving deeper into the subject of money.

In this second chapter, we will explain what saving and investing are and analyzing which choice is more convenient today.

Saving means taking out a portion of income received that you deliberately choose not to consume immediately but to store in a bank account for the future. Saving often results in the assurance guaranteed by the availability of resources when dealing with unexpected situations.

Savings can then be allocated to investment, and this is the main analogy between the two concepts. The investment may be of the "economic" type (such as the purchase of a car or a company machinery) or of the "financial" type (such as the purchase of a security or mutual fund with the objective to see capital growth over time). However, unlike savings, in the case of investing, the achievement of the desired objective is not certain (for example, a stock may lose value), so the result can be negative, compromising the amounts saved.

Which is better?

If the question is whether it is better to save or invest, the answer is probably "both". The choice depends on your financial situation and your personal goals.

Savings can be used to invest, but can also be used in other ways. In fact, the money saved can also be deposited in the bank to reduce risks (theft). But this, unlike what many thinks, is a wrong and unprofitable choice: money, in fact, tends to lose purchasing power over time due to inflation. In other words, if you save 100 Dollars today, in 20 years, you will be able to get less out of that money than today. This is why saving money is often the wrong choice if you want to get wealthy.

Assuming an average increase in the cost of living is around 2% and you have a saved sum of 5,000 Dollars. In five years, this sum will fall to 4,500 Dollars, which is 10% less (excluding banking taxes!) Obviously, you can keep the savings at home (under the classic mattress!) but with all the risks that come with it.

What is the difference between trading and saving?

Let's repeat it once again to get it better. Saving means to put money aside little by little in order to accumulate a certain sum. Usually, you save for a certain goal, like going on vacation, buying a car, or for emergencies that could happen.

Instead, trading means to take a part of the money to make it grow and buy tools that can increase its value like currencies, real estates, and ETF's.

Who should save?

Obviously, everyone should try to save a part of their money. The rule is to have something on your bank account that is necessary to "survive" for three months and cover the main expenses (such as food and rent). This will offer air pocket, in case of inconvenient and unexpected situations.

Saving is, therefore, a rule. And as every good rule, they have their own exceptions. You can stop putting aside the money when:

- You have too much debt and you are trying to pay it off;
- The family has priority and could not go on in case of unfortunate events to one of its members.

Even when you have set aside enough for emergencies, you do not have to stop saving. The goal of everyone should be to put aside at least 10% of their salary every month, perhaps starting from 5% and gradually scaling up. To make things easier, you can save money by thinking of any goals like having enough money for a great honeymoon or to get a new car.

Having a goal is essential so you know what you're saving for. Every rich person has financial goals, so it is a good habit to pick up.

When is the time to trade?

Like when you save money, you need to have a goal to when and how to trade your savings. In this case, it is important to know what your short, medium, and long-term goals are.

- With "short term," we mean goals for the next 3 years
- With "medium term," things are planned for the next 3-10 years
- The "long-term" goals are those for which you will not need the money back for at least 10 years or more

For short-term objectives, you usually invest through deposit accounts, which allow you to get a minimum return in a short amount of time. However, this has been a bit shrinking in the last period (deposit rates are at the lowest). For the medium-long-term objectives, it is, instead, advisable to invest in the market to avoid the reduction in value that inflation produces on "still" money. The market guarantees usually higher returns than deposit accounts over longer periods. Also, having a well-constructed portfolio helps a lot in this regard.

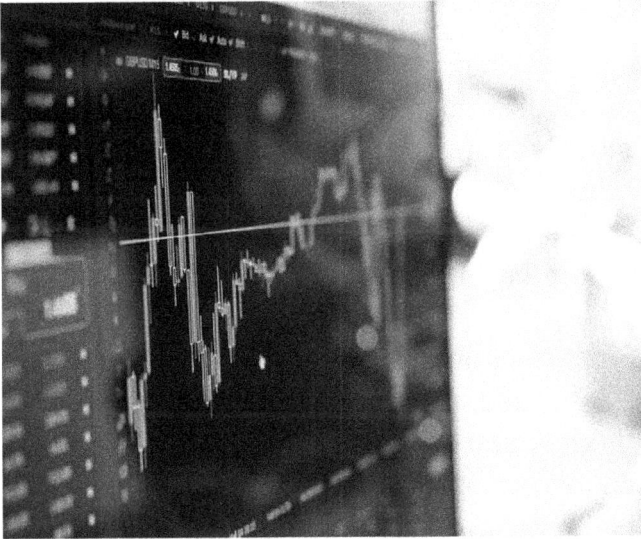

For those approaching or exceeding 30 years of age, having a medium-long-term goal is advisable. Investing and setting aside money for retirement can be a good start.

To sum up the concept, everything depends on your time horizon:

- If you think about using the money within one or three years, save it.
- If you do not need this money for the next 10 years, invest it.

If, on the other hand, you plan on using the savings in the next 5 or 10 years, but you want to still have money set aside in your bank account, then you will have to do both. Keep in mind that this is much harder and requires more discipline. However, with the right mindset, it is certainly the best option.

What does trading wisely mean?

Since the importance of the investment is well established, it should also be emphasized that there is no recipe to guarantee the success of an investment.

However, following some prudential rules can help minimize risks.

First of all, we need to avoid the dream of making money overnight. On the market, there are professional operators and experts who dedicate all their time to this activity, but they often make mistakes as well. This is just to say how difficult it is and how "get rich quick schemes" do not exist.

One strategy that every investor needs to master to reduce the risk is diversification. This means not putting all your eggs in one basket, but spreading your resources on different assets. When the invested amount grows, it becomes more important to diversify not only between the asset classes (stocks, bonds, and commodities) but also geographically (taking into account the currency variable) and size-wise (small or big cap companies stay within the equity, more or less, long maturities for government securities and bonds with different level of risk in the corporate sphere).

Making these choices takes time that needs to be subtracted from work or other activities. So, in the end, it is about investing time before moving the money. But it is worth it and, frankly speaking, it is the only option to avoid reckless choices that you may regret afterward.

Chapter 3: Money Excuses

Let's not hide behind the common opinion that you do not have money to invest. Do not get us wrong, you may be in the situation where money is tight and you do not have the resources to make a decisive move in the market. However, you can always control your cash flow and add extra streams of income. These will provide you with more money that you will need to save for future investments. We know it is hard, but it is possible and most millionaires started with nothing.

To us, the mindset is extremely valuable and, in this chapter, we want to debunk all the most recurring excuses people use to avoid or postpone their investments.

1. "I do not have time to trade."

One of the most common excuses is to believe that investing can take away most of the precious time we have available. The truth is that we are committing a big error of assessment. Trading does not require a specific amount of time: you can choose how much time you want to dedicate to it.

Obviously, the more the better, but you can even start with a few minutes a day.

Thanks to the advent of the internet and new technologies, investing is now just a click away, thus reducing not only the costs of negotiation but also the time required.

2. "I do not have enough money to trade."

To believe that investing is a subject reserved for those with large quantities of money is one of the worst mistakes we can make. Let's dispel this myth immediately. It is not true that to make money, we need big money. There are affordable financial products that do not require the fortunes of Scrooge McDuck to start planning your future.

From today, it is possible to start trading and investing starting from just 5 Euros.

Think about it, 5 Euros equals 5 coffees a week. If we had saved a coffee a day for 5 days a week since the Euros came into force until the end of 2016, we would have put aside a "small" sum of €3865. If these savings, instead of being forgotten in our piggy bank, were invested in global equity markets, at the end of

2016, we would have had €7493. No savings are, therefore, insignificant to be invested.

3. "I do not have the skills to trade."

One of the reasons that drive us away from investing is that we convince ourselves that we do not have the right skills and knowledge. Trading in the forex market may seem apparently difficult but the truth is that you do not have to be Warren Buffett to start doing it. By investing in mutual funds, for example, our savings are entrusted to a team of expert managers who make the investment choices for us on a daily basis. While we let others manage our money, it is fundamental to learn. Remember that the goal is to become an investor that takes care of his resources.

4. "I will trade in a few years when I have a higher salary"

Delaying an investment is not a wise choice, especially considering the benefits of compound interest capitalization. To show you, we have compared two capital accumulation plans: the first invest a sum of € 100 a month from the age of 25, while the second invests € 200 a month but starts from 35 years.

In your opinion, which of the two would be able to obtain a higher capital at the age of 70 years old?

The accumulation plan of €100 monthly undertaken for 25 years will have generated at the age of 70 years old a capital of 520,000 Euros: 50,000 more than the other.

Anticipating the investment not only requires less economic effort but also allows you to obtain higher earnings compared to a higher investment delayed over time.

5. "Trading is too risky."

None of us wants to lose money, but we do not realize that we are already doing this when we decide not to trade. If the alternative to investing is, in fact, to feel safe by parking our savings on the bank account, inflation could reserve us unpleasant surprises reducing inexorably our purchasing power in the future.

If you trade using portfolio diversification and adopting a long-term time horizon, the chances of losing money are reduced a lot.

Chapter 4: It is a Question of Money

When it comes to trading strategies, the amount invested cannot be ignored.

This chapter is oriented to the management of assets between 10,000 and one million Euros. Another premise for reading is to have a clear idea of what is meant by the amount investable.

We will divide our field of action into three bands. All three bands will assume that it has already:

- trade the maximum tax-deductible share in the supplementary pension;
- stipulate any life insurance; indicates that all the negative points described in Life Insurance should be considered;

- deduct from the tradable portion any allowances for false investments, i.e. secondary activities that are

22

genuine alternative works.

The last point is particularly important for investments in real estate and land. As we will see in the operational plans, for assets up to € 250,000, a speech on property and land can only be marginal.

The main reasons for the previous statement are:

1. Such trades often tend not to be real.

In the modern sense, an investment is if it requires a minimum allocation of resources (for example, I buy 10,000 Euros of government bonds); otherwise, it is configured as a real activity.

Buying a home that you then rent is the simplest example.

If we interact directly with the tenant, we are doing a real activity, an alternative to our work in which we often do not take into account the management costs and the time we spend. This is different where we limit ourselves to buying the house and entrust to a paid external structure the role of the administrator of the building. In this second case, what remains is the real gain of the rent. The same applies when buying agricultural land: only by considering it an activity (i.e. cultivating it and managing it with appropriate decisions) will we be able to make the most of it.

2. These trades minimize management costs only for large capital.

In fact, the realized capital gains are the gross of the taxes and of all the management expenses that serve to maintain the asset in question over the years. For small investments (for example a house worth 300,000 Euros), the inflation, taxes, maintenance costs, etc. reduce the real gain considerably.

Trading instruments

Below are the trading tools that can be considered:

1. properties and land;
2. instruments for maximum liquidity (i.e. can be liquidated in up to 3 months);
3. bonds;
4. stocks;
5. Currencies (Forex)

The individual instruments must then be optimized by following the instructions given in the following paragraphs.

We must warn against investing in alternative and typically speculative fields (art, jewelry, etc.) without having a specific capacity. These fields are, in fact, similar to alternative work: buying a painting, a prestigious watch, or a classic car hoping for a great revaluation is completely optimistic if you are not an expert in the sector. On the other hand, if one is, it makes no sense to make it all occasional, but it would make sense to make it at least a second activity.

The proposed management is mainly passive, in the sense that we must follow the trend of our trades not continuously over time, but with periodic checks (for example, quarterly) to verify whether it is appropriate to positively disinvest. For example, if a currency was bought a year ago at 95.25 and is now worth 99, a 4% gain justifies the sale; if, on the contrary, it has fallen to 94.20, it will put the heart in peace and will be held until its expiration.

From 10,000 to 50,000 euros

I know I am disappointed by those who thought of diversifying, but with such a modest sum, you can only use two tools: forex and bonds. You can use together or, better, use the latter unless the former is no longer advantageous due to a particular economic situation.

From 50,000 to 250,000 euros

Here, the four instruments are all usable but obviously, still with due consideration.

For buildings and land, it is advisable to include them in the additional quota. If you decide to invest 50,000 Euros in real estate instead of buying a tiny studio, it makes more sense to buy a bigger house of ownership. The fees on the added quota are less than on a second home, and there would not be all the hassles of managing an asset which, due to its small size, would yield modest yields in any case of a certain management commitment.

Also, in this case, the bonds take the largest share of the investable amount (at least 50%) and can be replaced by the instruments for maximum liquidity only in exceptional cases in which they make more.

The actions deserve separate speech. In theory, with a capital of € 250,000, it would be possible to invest in the shareholder. However, in practice, it is better to do so by linking the figure to one's age.

If at 30 years old, an invested share of 40% can be significant, at the age of 60, it should not exceed 10%. With these data, it is automatic to remember that at the age of 40, a maximum of 30% is invested and at a maximum of 20%. I propose you the rule of 70. The sum between the age and shareholding always makes 70.

Let us remember, however, that investing in the stock is an opportunity, not an obligation.

From 250,000 to one million euros

We are now on important figures. Before going into detail, it is necessary to understand "what wind it pulls". Currently, with an economy still in partial crisis, it seems that the situation is this:

1. secure bonds and liquidity: ****
2. actions: **
3. gold: **
4. properties: *

This picture will appear disappointing to those who dream of speculating with their capital, but it is certainly the one that protects it most.

For those people, forex might not be the best option.

With regard to property and land, up to 30% of assets can be invested in them both as an additional share and as an investment in its own right. Many would come to invest up to 100%, but it is a too simplistic solution because, in fact, if you want to invest in the brick with such capital, it makes more sense to undertake a real second activity. Furthermore, it should be remembered that *a property has value only if you can resell it!*

In other words, instead of investing in a couple of luxury apartments in the city center, it is more logical to invest in smaller units by diversifying the risks that are always present on the individual investment. In any case, the crisis in the real estate sector that began in 2008 has, in fact, extinguished optimism that lasted for decades, optimism without a real rational motivation.

Once the portion allocated to property and land has been determined, the amount to be invested in shares must be determined. Also, in this case, the maximum is represented by the rule of 70. The remaining part is destined for the bonds.

Chapter 5: An Introduction to Forex and the Difference between Trading and Investing

Most beginner investors and traders have quite confused ideas when approaching the forex market, trading currencies (or options, ETFs, commodities, etc.), or trading in general.

One of the pivotal points that create confusion in the mind of those interested in making their money work through investments is the lack of understanding of the crucial difference that exists between trading and investing.

The confusion derives from the fact that in the eyes of the investor or the uneducated and non-conscious trader, doing trading or investing seems to be the same thing.

In reality, although they are united by the desire to make a profit, the two operations arise from different logics and follow different rules.

In fact, those who invest in a measure of the value of what they buy (an action, a house, a business, an object of art, etc.), try to buy it at a discounted or otherwise balanced price, and the entire operation is based on the prediction or hope that, over time, the good purchased will increase in value and that this increase in value will automatically be reflected in a corresponding increase in its market price allowing it to be sold for a profit.

An easily understandable example of investment is those who buy agricultural land in the expectation that it will then be buildable.

The greatest investors of history, such as the legendary Warren Buffett, are in fact masters in buying "depreciated quality". Of course, their time horizon is never very short and the value of what they have purchased can remain or even go down for a certain period of time without this causing them to worry excessively.

Who trades, however, does not bet on a change in the value of things. To be honest, the hard and pure forex trader does not care highly about the objective quality or the nature of what he buys, he is only interested in acquiring it at a price that (in a generally rather short time frame) he plans to grow, regardless of the fact that the value of what he purchased remains perfectly identical.

In fact, what makes trading possible is simply the fact that the prices of things (and, therefore, also investment objects such as shares, bonds, real estate, etc.) may vary regardless of their value due to the law of the application and of the offer.

An example out of context of activity comparable to trading is that of the Super Bowl tickets reseller, who obtains the tickets three weeks earlier at regular prices and then resells them at the last moment when tickets are already sold out and he can market them to a much higher price.

On an exchange, for an investor, it is crucial to understand what he is buying and what the current and future value of the company he is planning to buy shares. In other words, investors search quality companies that are currently depreciated.

On the other hand, for a forex trader, it is sufficient to use tools (generally, the stock's graph evaluated through technical analysis) that allow him to make a forecast of the future price of the stock regardless of the value of the company and its corporate purpose.

Chapter 6: What you Need to Start Trading in the Forex Market

To start trading, you must meet the following requirements:

- a PC with a stable internet connection;
- an online trading platform, to be chosen among those recommended and regulated by us;
- all the recommended brokers offer adequate training to all traders whether they are novice traders or experts;
- graphs relating to market quotes in real time;
- economic news;
- comments and operational suggestions;
- A great desire to learn.

Today, thanks to online trading, it will be possible to obtain profits that can reach up to 70% based on the chosen broker and a fair trading preparation, without risks of losing the entire capital invested. Therefore, we advise you to apply scrupulously and follow all the advice that your broker provides you.

In addition, thanks to the many materials that can be found in this book, you can start learning what the right terms of online trading, how you can invest in the forex market, etc.

When investing in the forex market, you can do mainly two different types of transactions:

- Long operations (upward investment)
- Short transactions (downward investment)

In other words, when you are trading, you can buy currencies and sell currencies. The goal, however, remains the same: to make a profit. When you want to buy, you will only get a profit if the value of the currency will be increased when we want to sell it. For example, we buy 100 lots of a currency when it is worth $1.1 each, and then we sell them when they're worth 1.2$ each. The price difference is multiplied by the number of lots equals our profit.

When you want to sell, it becomes a bit more complicated. In short transactions, in fact, it is the broker that lends us the number of shares on which we want to invest on the downside. For example, the broker can lend us 1000 Euros listed at $1.23 each. The securities that the broker lends us for a short transaction are sold immediately. The profit remains "frozen" in their trading account.

This profit will be used to buy back the same amount of currency that the broker had lent us because we have to return these currencies to the broker. In that case, we will have a gain if the value of the currency has fallen.

The difference between the initial sale of the securities lent and the expense to repurchase them is +€500 in this case. That is our profit. If, on the other hand, the value of the currency increases, we will have to spend more money than those earned from the initial sale of the pre-arranged securities. In this case, we will suffer a loss.

Chapter 7: Not only Forex

The main method for investing in the forex market, therefore, remains the classic forex market. When you operate on the forex market, you are actually buying and selling currencies.

However, over the years, other financial instruments have been introduced to invest in forex and currencies indices on the forex exchange. We are talking about CFD (contract for difference) and binary options. The main feature of these two financial instruments is the following: when you use them to invest in forex, you will not actually own the lots you are investing in.

That said, for those who do not intend to trade online, it could make little sense. Let's try to clarify. Both CFDs and binary options are contracts between investors and brokers. It's not like the classic forex market where traders buy and sell among themselves. In CFDs and binary options, the asset movement (in this case, the buying and selling of currencies) does not take place.

CFDs and binary options are used to speculate

on the performance of the value of equity securities. If the trader's forecast is correct, the operation will lead to a profit. If the trader's prediction is wrong, the operation will lead to a loss. So the mode of operation is similar to the stock market: if I invest on the upside, whether I do it with CFDs or actually buy currencies, I only earn money if the value increases.

As we explained in the previous paragraphs, CFDs are also derivative instruments, so they are used to speculate on the performance of asset values. This means that when you buy and sell CFDs, you will never own the asset traded (as opposed to classic forex trading).

Moreover, as with binary options, with CFDs, it is possible to trade on:

1.   Equity securities
2.   Equity indices
3.   Forex currencies pairs
4.   Commodities
5.   ETF

Leverage plays an important role in CFD trading: through leverage, we can literally multiply the value of our investment. Just to give an example, if you use a lever of 1: 100 and invest €100, you can move well €10,000 (using only your hundred!). All this is made possible thanks to the leverage, which is a sort of "loan" (if we can define it) by the broker, which you can invest more money than you really have.

This means that your earnings, but also your losses, will be calculated not on the €100 you really invested, but on the € 10,000, you will have invested thanks to leverage. Therefore, the lever can amplify the gains, but also the losses. To see an example of CFD trading, we refer you to our article on how to trade CFDs in equities with eToro.
eToro is one of the leading CFD brokers, very suitable also for non-professional traders and those who want to approach the world of online trading (thanks to the free and unlimited demo account offer).

But if we talk about eToro, we can't avoid talking about Social Trading. For those who do not know, eToro was the first broker to have introduced Social Trading in CFDs. Thanks to social trading, it is now possible to invest by copying (automatically) the operations carried out by the other traders registered on the eToro platform. All you need is a couple of clicks to find the traders to follow, choose the amount to invest, and you're done. In this way, even novice traders can exploit the knowledge and experience of professional traders by copying their operations.

The online trading strategies are based on the study of mathematical and graphic analysis that can suggest the trader the best moment to buy and sell. As we have seen today, it is possible to invest in the stock market thanks to online trading, choosing between trading binary options and trading with the forex market.

However, there is no suitable trading strategy for all traders, but there are different trading strategies based on the traders and their style of trading. Therefore, it is possible to customize different online trading strategies on the basis of their trading objectives, their intellectual, and psychological abilities.

We also recommend using 2 proven techniques to not turn winnings into losses:

> 1. Stop loss: it establishes a maximum loss that you are willing to suffer;
> 2. Take profit: you place a dynamic

40

exit level that rises slowly.

Chapter 8: How Much Money to Start With?

Many people ask themselves this question: how much money do they need to invest in the forex market? An entirely legitimate question, but whose answer varies according to many factors. First of all, consider how you have decided to invest in currencies. The classic forex market requires capital of a certain size, usually (at least) from €5,000 - €10,000 to start investing in the forex market.

If you do not have these figures or do not feel ready to invest them, you can use other financial instruments such as CFDs and binary options. In both cases, the minimum capital to invest is really limited. We usually speak of €100-200 to open an online trading account and have access to a trading platform to invest. Obviously, no one forbids you to start investing a larger amount. Our advice, especially if you have never done online trading before, is to invest a maximum of €1000 in your first CFD trading account or binary options.

Regardless of how you have decided to invest in the forex market, however, remember to choose only figures that you can afford to lose. What does it mean? It means that you have to invest in figures that will not put your financial stability at risk.

Another topic that must always be dealt with when it comes to investing in the forex market is the risk. Let's clarify it. Trading online or investing in the forex market is risky. The risk is a factor that cannot be eliminated. Anyone who tells you that online trading and investing in the stock market is risk-free and easy is lying shamelessly. Trading online is risky, as it can result in the loss of your capital (in case of bad decisions).

Management-Money-and-Risk

All investment activities are risky. Whether it is the real estate market, opening a business or starting a start-up, or online trading, the risk is always present. The important thing is to know how to manage the risk factor so that it can be reduced and controlled. Nevertheless, remember that you will never be able to completely eliminate the risk factor. It will always be present in all of your future forex exchange transactions.

This means that sooner or later, you will lose

money by investing in the forex market. After all, no one is perfect. Being able to make only profitable investments is impossible. We must accept the fact that suffering losses "is part of the game". But the important thing is to earn more than what you lose. If, for example, out of 10 investment transactions you miss 2 but you earn from the remaining 8, you can say that you have reached an excellent goal (and profit).

Fix yourself with the idea of not wanting to lose money by investing in the stock market. It will not help if you want to start trading online. Professional investors know very well that losses must be learned to accept, and aim to limit their number (check the term well: limit, do not eliminate). In this way, by limiting losses, profits will increase, which is the goal that every trader should operate on the forex exchange.

What is the minimum capital requested to invest in the stock market? In this chapter, we will try to answer this question, with attention to the type of operation and market. First of all, we specify that when we talk about the minimum capital to invest in the stock market, we are talking about something very different from the minimum capital to invest in Forex.

In fact, while the investment in Forex has an essentially speculative activity, in the stock exchange investment, dividends, coupons, and long-term investments are also to be assessed. Now, let's analyze how much it takes to invest in the stock market.

As mentioned, the minimum capital to invest in the stock market depends mainly on the type of operation that you intend to have. We can say with absolute certainty that the smaller the duration of the investment is, the lesser is the required capital.

If we operate with a maximum daily duration, it is presumable that every evening, we close all open positions so our minimum capital to invest in the stock market, even without leverage, may even be only €1,000 allowing at least one transaction per day. Operating with leveraging the minimum capital to invest in the stock exchange may be lower, even a few Euros, if we use a high leverage. It seems obvious that if the duration of our investment will be semi-annual, the minimum capital to invest in the stock market will have to be decidedly higher so as not to be in the sad condition of being able to operate once every six months.

So far, we have talked of minimum capital to invest in the stock market exclusively from the point of view of the number of possible transactions. However, it is not so simple. The minimum capital to invest in the stock market should be sufficient to diversify our exposure to avoid large losses. The ideal situation would be to invest in at least three markets with only two market transactions. It goes without saying that with a very short duration of leverage operation, the minimum capital to invest in the stock market will be from a few thousand euros, while for half-year transactions, the capital will necessarily have to be higher.

So far, we have talked about optimal operations. The minimum capital to invest in the stock market will depend very much on your risk profile. The greater your risk appetite, the higher your leverage will be and, therefore, you will be able to invest with a smaller capital.

Chapter 9: Compound interest and Forex

Earning more means being paid more. We usually think that others should pay us more if we want to make more money. But this is not always true. We can earn more even if we pay ourselves more, and not the others.

This is a fundamental principle underlying the financial success, first disclosed in 1926 by George Samuel Clason through his book entitled *The Richest Man in Babylon*, a great motivational classic.

The principle states that part of what you earn must be maintained. Putting aside at least 10% of what you earn and making that money inaccessible to ordinary expenses and possibly even extraordinary expenses. You can increase this amount exponentially over time. Considering any investments, thanks to the power of the compound investment, the amount saved/invested over the years can become important. In fact, many people are able to earn more and build their assets by paying themselves first. It is a true and effective principle today as it was in 1926.

Yet, as this 10% formula is easy, people are unwilling to listen to it and apply it. This is because you are usually looking for tricks to get rich quickly, and you do not have a medium to long-term vision. On the other hand, having a long-term investment plan is a solid foundation on which to build one's own economic stability. And you can start earning more by paying yourself first from today. The earlier you start, the quicker you will build your financial success.

Using the Power of Compound Interest

To earn more, you can take advantage of the compound interest. Here's how it works. If you invest 1,000 Euros at a 5% interest, you will earn 50 Euros of interest and, at the end of the first year, you will have a total investment of 1,050 Euros. If you leave both the initial investment and the interest earned on the current account, you will receive a 5% interest the following year over €1,050, or € 52.50. In the third year, you will earn 5% out of 1.102.50, and so on. At this rate, within 15-30 years, your money will turn into an amount well above the sum invested initially. But precisely how much does the invested capital grow? The Italian mathematician Luca Pacioli explained it in the fifteenth century: any capital doubles in a number of years equal to 72 divided by the interest rate. Returning to our example: if the interest is at 5% per year, we divide 72 by 5; which make 14.4, i.e. in 14 years and 4 months the initial capital doubles. The sooner you start the bigger the result will be, as you will have more time for the interest you capitalize to produce its powerful magic. Start now to save and invest for your future, even if you do not have a large sum. You do not need to have an extra sum of money. You can start with any amount and grow it over time.

The Secret of Paying Yourself First

If you want to earn more money by paying

yourself first, you have to make savings and investment a central part of your financial management, just like the mortgage payment. Get accustomed to saving a fixed percentage (at least 10%) of your monthly income and investing it in special savings account that you decide not to touch. Ideally, this step would be automatic, such as a fixed monthly deduction on your paycheck. The automation will ensure that you will not have to rely on your self-discipline and your ability to save will not be affected by your mood, from domestic emergencies or otherwise. Continue to increase that account until you have saved enough to invest the sum accumulated in bonds, in a mutual fund or in real estate (spending money on rent without building any assets is really a waste). Let your investments build your assets over time, and try to live with what remains after you have paid yourself. If you want to spend, try to earn more to afford it. But never put your hands on your savings to finance a more ambitious lifestyle. The ideal would be for your investments to grow to the point where you could live with interest, if necessary. Only then will you really be financially autonomous and free.

If you want to earn more, you need to create assets, not liabilities. Rather than spending all the money you earn by enriching someone else, invest in assets that produce other income (stocks, bonds, real estate, gold, etc.). Then, when your money starts to grow, educate yourself further about the best way to invest your money. Stay informed on news about investment opportunities and remember to protect what is yours through a good insurance policy. Do not blindly trust who will manage your money, but always try to improve your financial education. This will make you a financially prepared person ready to get rich. Once you understand this, money will follow.

What is a compound interest? Not everyone may know how to respond immediately to this question. In fact, if everyone knows what the simple interest is, i.e. the one that withdraws at the end of the agreed time unit, fewer are those who will know what the compound interest is, how it works and, most importantly, how to take advantage of it.

The Example of a Bank Account is Enlightening.

If on 1 January I have a net rate of 1% on my account, at the end of the year, I have 101

Euros. The euro is added to the capital and, if the conditions do not change, at the end of the second year, I will not have 102 euros, but 102 euros and 1 cent where the cent represents 1% of the euro accumulated after the first year.

So far, everything is clear, but most of us cannot calculate the compound interest of an investment and tend to treat it as simple interest. This is due to its slow start, especially with small capital, that tends to be treated as "irrelevant". However, there is nothing more wrong that an investor could do.

If, for example, after 5 years of investment, my capital of 100 euros is now 140, we are led to believe that the interest was 8% per year.

This is incorrect because, in doing so; we do not take into account that at the end of each period, the interest accumulated has gone to increase the capital. If the interest had really been 8%, composing the 5 years we would have had:

Initial capital: 100

- 1st year: 108

example).

In this way, Marie has a greater incentive to have to delay her purchase.

When Julie returns the sum loaned, she will receive €1,050 instead of € 1,000.

The following month, Marie can then buy the air conditioner and, to celebrate, use the €50 interest to go out to dinner with her boyfriend.

In short, in the end, this recognition for the delayed use was not bad!

Now that we understand the concept behind the rate of interest, it is good to enter a little more in detail and make some distinctions.

In this regard, we can divide the interest rate into two broad categories:

1. The simple interest;
2. The compound interest.

Simple Interest

Let's go back to the previous example.

At the end of the period, Julie returns the money plus the interest to Mary. Soon after, however, the girl asks again the same amount to buy a new refrigerator, as the old one suddenly broke.

Marie agrees to lend the money back to her friend.

The following month Julie firmed up her debt plus new interests, again for a total of € 1,050.

Now Marie is with her initial capital plus €100 in interest for a total of € 1,100.

Interest is defined as simple when, once it has matured on the underlying capital, it does not generate further interest.

In our example, we note that the first €50 was not added to the capital loaned the second time.

Compound Interest

Julie asks Marie to lend her €1,000 with the promise to return them in two years. Mary agrees, as long as Julie accepts a compound interest on the mature borrowed capital.

In this case, Julie will not have to pay the interest immediately at the end of the 1st year but will add the €50 interest in the capital, which in turn will accumulate 5% in the 2nd year.

At the end of the agreed period, Julie must return:

1.   €1,000 capital
2.   €50 interest for the first year (€ 1,000 + 5%)
3.   €52.50 interest for the 2nd year (€ 1.050 + 5%)

The total capital to be returned to Mary is € 1,102.50.

Here we have materialized €2.50 more than the previous example, due to the compound interest.

The interest is defined as compound when, once it has matured on the underlying capital, it is added to the latter and contributes to generating further increased interest in the future.

Do you understand why the compound interest is your new best friend?

When you deposit your money in the bank account you are doing as Marie, that is, you are "lending" your money to the bank, which uses them to perform its credit function and lend it to people and businesses.

As a reward for this service, you are given an interest in the sums deposited, that is, a reward for the fact that you delay their use.

How to take advantage of the compound interest

If you do not want inflation to eat a nice slice of the real value and the purchasing power of your money, you have to make sure that the latter accrue compound interest over time.

Certainly, a part of the liquidity at your disposal, you can deposit on one or more deposit accounts or accounts with limited operations where higher interest rates are recognized.

For example, you could deposit your emergency fund.

The rest, however, should be invested in a portfolio of efficient financial instruments that protect your capital and create added value.

The compound interest must be exploited for at least two reasons:

1.   Increase savings while waiting for their use;
2.   Defense against inflation.

A wise thing to do is to exploit the power of compound interest to make the value of your money grow faster, protecting it from loss of purchasing power.

Try to keep only small amounts on bank accounts that give you little to nothing.

You can leave just the right liquidity for your daily expenses and for the emergency fund.

Chapter 10: Seven Things to Consider when Trading Forex

It is not true that to invest in forex, you have to have a lot of money. It is true that with equal choices, a successful investment makes it proportionate to the money invested. But the opposite is also true, namely that if you make a mistake with so much money, you lose more than if you make mistakes with a few. But the really important thing is that, with whatever sum you start, the stock exchange can give you earning opportunities. It all depends on how you invest. It is not an easy thing, as long as it is said, and it takes time and attention. There are two ways to earn from the forex market: cash out the dividends distributed periodically by the currencies you have invested in, and sell your currencies at a higher price than the one you bought them. In short, in order to trade in currencies and get a profit, one must know how to choose.

Here are 7 things to keep in mind.

1. Plan the investment

The first advice that we can give you about

financial investments is about planning of investments or understanding the best actions to buy and diversify your portfolio.

Even if you have never experienced this first hand, it's not a problem. You will learn about it sooner or later.

In order to better diversify your currency portfolio and understand where to invest, we recommend opening a demo account.

The demo account allows you to not only to plan investments but also to:

- Carefully analyze the stock market on which you want to invest;
- Plan your investment strategies;
- Familiarize yourself with the platform;
- Get familiar with the market.

If you decide to buy shares in an unconscious manner and then open a real account and invest without the right measure, then prepare to say goodbye to your immense capital.

Of course, this is not the most appropriate and wise way to invest.

2. Draw the investment plan you just made

To quote W. Edwards Deming, world-renowned essayist and quality management consultant:

"If you cannot describe the process of what you're doing, you do not know what you're doing."

As for everything that requires a certain discipline, it is important to outline its investment strategy. In this way, it will be easier to articulate it. Once your strategy is written, look at it to make sure it meets your long-term investment goals.

Writing and schematizing your strategy will give you a firm base to start again in times of chaos and will make you avoid making important trading decisions dictated by emotions.

It offers you a clear outline to review and change if, with time and experience, you notice defects or if you change your investment goals.

If you are a professional investor, having a written strategy in black and white will help your clients better understand the investment process you are proposing.

3. Learn the difference between investing and speculating

Understanding the difference between a trader and a speculator is very important. You need to know how to "use" the difference if you want to make the most out of your investments.

Before buying currencies you have to evaluate:

- what do you want to get from the markets;
- what is your personal level of risk tolerance;
- if you are investing;
- if your goal is to speculate on the markets;
- the time you have available to spend on investments.

If you want to get the maximum profit in a tight time, then you must have a considerable minimum time to devote to the study of markets and financial instruments. So, you must understand the difference between speculator and investor.

What does a speculator do?

The speculator is the trader who buys and sells shares in order to make a profit in the short term. In this case, we are talking about very narrow trade times ranging from a few minutes to a few weeks.

We do not talk about years or months.

They only take advantage of the price difference between the value of the sale and purchase of the deal.

The speculator's characteristic is that it is not interested in dividends distributed by listed companies.

What does a trader do?

Contrary to the previous one, the investor, also defined as a long-term investor, invests his capital by providing liquidity to the currency pair.

In this case, the trader will buy the so-called "lots" of a given currency. What is the goal of a trader? It is to keep the stocks in his wallet for a prolonged period of time and make them profit!

This allows him to benefit from the detachment of the dividend that is added to the possible appreciation of the title.
It is very important to understand what kind of investor you are. Pay close attention to this step because it is essential in earning with investments in the forex market. Most of the trader's operating strategies are based on fundamental analysis that is very different from those of a short-term investor or speculator.

4. Understand the importance of timing (and the impossibility of getting it right)

It is very important to understand when the right time is to buy and sell currencies.

In this case, the timing is an indispensable part to identify the stocks to be bought. If the correct price levels are not identified, there could very well be the risk of entering the market at a risky point. This could be unfavorable and does not allow us to accurately quantify the transaction's risk-return ratio.

5. Learn your strength and weaknesses

Does your investment strategy follow your idea of how investments depreciate or appreciate? If so, how do you exploit your knowledge?

This question refers to your actual knowledge of the market. Ask yourself:

"What makes me smarter than the market? What is my competitive advantage?"

You may have special knowledge of the industry or have access to a study that few others know. Or, you could get your own opinion by exploiting some market anomalies, as what happens in the strategies for the purchase of securities with a low price/value ratio.

Once you have decided what your competitive advantage is, you need to decide how you can use it profitably to develop a trading plan.

Your investment plan should include rules for both purchase orders and sales orders. Also, keep in mind that competitive advantage could lose its profitability and its effectiveness if other investors will begin to adopt your own investment strategy.

Or, you can be convinced that markets are totally efficient, which means that no investor will ever have a real competitive advantage. In this case, it is better to focus on minimizing commissions and transaction costs by investing in passive instruments such as futures.

6. Is your strategy versatile?

There is an old way of saying on Wall Street:

"The market can remain irrational longer than you can remain solvent."

Successful investors know where their investment performance comes from and are able to explain the strengths and weaknesses of their strategy. As trends and economic issues change, many investment strategies have periods of great performance followed by periods of poor performance.

Having a good understanding of the weaknesses of your investment strategy is essential for maintaining confidence in the market and investing with conviction, even if the strategy is temporary "out of fashion".

7. Understand that a good strategy can be measured

It is difficult to improve or fully understand something that cannot be measured.

For this reason, you should always have a benchmark to measure the effectiveness of the investment strategy you are using. This benchmark must be consistent with investment objectives, which in turn, must tune into your strategy.

There are two types: the relative benchmark and the absolute benchmark. An example of a relative benchmark could be the EURUSD pair. An example of an absolute benchmark could be a performance target.

Even if it is a time-consuming process, it is important to consider the amount of risk you are taking with respect to the investment benchmark. This can be done by recording the volatility of portfolio returns and comparing it with the volatility of benchmark returns over certain time periods.

Chapter 11: Forex and Leverage

What is leverage?

Through the use of financial leverage, a person has the possibility to buy or sell financial assets for an amount higher than the capital held and, consequently, to benefit from a higher potential return than that deriving from a direct investment in the underlying and, conversely, to expose himself to the risk of very significant losses.

How does the leverage work?

Let's see how the concept of leverage works starting from a simple case. Let's assume you have €100 available to invest in a currency pair. Let's assume that the gain or loss expectations are equal to 30%. If things go well, we will have €130. Otherwise, we will have €70. This is a simple speculation in which we bet on a particular event.

In case we decide to risk more for our investment in addition to our €100, and another €900 is borrowed, then the investment would take a different articulation because we use a leverage of 10 to 1 (we invest €1000 having a capital initial of only €100). If things go well and the stock goes up to 30%, we will receive €1300. We return the €900 borrowed with a gain of €300 on an initial capital of €100. So, we get a 300% profit with a stock that only gave a 30% return. Obviously, on the €900 borrowed, we will have to pay an interest, but the general principle remains valid. The leverage makes it possible to increase the possible gains.

Consider the further case of the investment in derivatives. Let's assume we buy a future that, within a month, gives the right to buy 100 grams of gold at a price set today of €5,000. We could physically buy the gold with an outlay of €5000 and keep it waiting for the price to rise and then sell it back. If we decide to use derivatives instead, we should not have €5,000, but only the capital needed to buy the derivative. Let's say that a bank sells for €100 the derivative that allows us to buy the same 100 grams of gold in a month to €5,000. If in a month the gold is worth €5,500, we can buy it and sell it immediately, realizing a gain of €500. With the €100 of the price of the derivative, we make a profit of € 400, or 400% with € 100.

This is how leverage works. Do you get the amazing power it can give to the average investor?

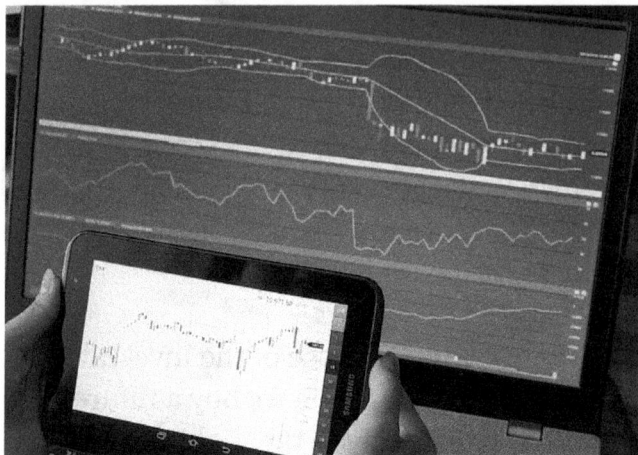

What are the potentials of its use?

The potential of leveraging is clear. But be careful. The leverage multiplier effect, described with the previous examples, works even if the investment goes wrong. For example, if we decide to invest €100 in our possession plus an additional sum of €900 borrowed, if the currency pair depreciated by 30%, we would remain with only €700 in hand; having to return the €900 borrowed plus interest and considering the €100 of our initial investment, we would have a loss of over €300 on an initial capital of €100. As a percentage, the loss would be 300% against a reduction in the value of the pair of 30%.

Another element to keep in mind is that different financial levers can be combined. In this way, speculation operations are carried out using a "squared lever" with clear reflections on potential potentials.

What are the risks related to leverage?

What may appear to be an interesting tool with positive potential for the investor, on the other hand, presents risks that must, therefore, be taken into due consideration. In fact, if the financial system, as a whole, works with a very high leverage and financial institutions lend money to each other to multiply the possible profits, the loss of an individual investor can trigger a domino effect by infecting the entire financial market.

Banks are typically entities that operate with a more or less high degree of leverage: against a certain net capital, the total assets in which the resources are invested are generally much higher. For example, a bank with equity of €100 and leverage of 20 manages assets for

€2,000. A loss of 1% of the assets involves the loss of 20% of the equity capital.

The development of the market for the transfer of credit risk (from financial intermediaries to the market) has meant that the traditional bank model, called "originate-and-hold" ("create and hold": the bank that provided the loan it remains in the balance sheet until maturity), has been substituted for many operators from the "originate-to-distribute" ("create and distribute": the intermediary selects the debtors, but then transfers the loan to others, recovering the liquidity and the regulatory capital previously committed or the pure credit risk, with benefits only on capital requirements), with the effect of a further increase in leverage. The spread of this second bank model is one of the factors that explain the crisis triggered on the sub-prime mortgage market.

Property price inflation has supported the issuance of loans and the exponential growth of the related market, allowing banks to make huge profits and, at the same time, increase leverage. But "the money machine" could not last long and, in the end, many banks found themselves without sufficient capital to absorb the losses deriving from the inversion of the real estate market trend, resulting in failed companies.

In the meantime, the example of the banks has spread within the financial system by spreading to all other financial institutions: leverage had prevailed, especially in the United States, generating a huge volume of risky investments that rested on a fraction infinitesimal of equity capital. We are thinking of the issue of so-called "credit default swaps" (derivative instruments used to hedge against the default risk of the debtor): some insurance companies were heavily exposed to the real estate market and when the latter collapsed and the value of mortgages fell, they began to lose without having sufficient capital to absorb the losses deriving from the issue of those instruments.

In order not to risk failing and return to sufficient levels of bank capital, capital increases (not an easy task in times of crisis), the reduction of the amount of loans to businesses (granting a lower number of new loans and not renewal of those already issued) and the disposal of other liquid assets (mostly shares) can be used. The result of all this, in the period of the sub-prime crisis, was a credit freeze and a collapse of the stock market.

These are the main channels through which the financial crisis has hit the real economy. Credit rationing has affected investments and the decline in the stock market (which adds to the decline in house prices) has reduced the value of household wealth and consumption.

We know that a certain level of leverage is physiological to sustain economic growth, even if we have no indication of what the optimal level is. But history teaches us how in an increasingly globalized and interdependent economic-financial system, leverage can be a trigger for speculative bubbles. And it is in these periods that the strongest disconnect between finance and the real economy is generated.

Chapter 12: A Different Currency Pair: Cryptocurrencies

One of the strategies that I am going to explain is trading in cryptocurrency. Why do I invest mainly in crypto and blockchain related assets? Because I truly believe they are one of the biggest revolutions undergoing in this very moment, and that this is the perfect time to get involved before the market explodes to the upside and prices rise at major stocks level.

Another reason that I like cryptocurrencies and their market is that they are extremely volatile and provide the average Joe the possibility to make serious money without investing a lot. It is not a secret, in fact, that every time the market starts to rise, people rush into the search for the "next big win" and the question that circulates is always the same: "What will the next cryptocurrency be that will go to 'the moon'?"

The issue with cryptocurrencies is that being a market that is not yet regulated in several countries, the risk of pumps and dumps, manipulation and fraud is just around the corner. This is why I wanted to cover them in this book. In fact, since they provide a great opportunity, I am worried that a lot of people may get involved without knowing what they are doing and will lose a lot of money down the line. Here, I want to show you what I do before investing in a particular asset and how I keep it a sustainable source of passive income.

Before getting started, here is a list of useful tools for the analysis of cryptocurrencies:

1. Coincheckup.com - one of my favorite sites, offers much more data than other cryptocurrency monitoring sites;

2. Coinmarketcap.com - one of the oldest crypto price tracking sites, far more popular than Coincheckup, but offers less data;

3. Blockfolio - another popular cryptocurrency tracker.

Now let's get to the good stuff.

Step 1 - Understanding your risk profile

Many people will advise you to buy "low capitalization" cryptocurrencies and tokens (i.e. between 10 and 100 million dollars) because they have a greater opportunity for growth in terms of percentage.

Although this statement is relatively correct, you have to keep in mind that the smaller a coin is the riskier it is to invest in it. Why? Because the project has indeed have a much higher risk of failing.

In traditional investments, most people aim and are happy to get an annual return of 3% - 4%. But they could be in serious financial difficulty if the invested capital is lost. So most of the time, more well-known, safer, and more stable titles are selected.

Other people would be satisfied only with an annual yield of 7% - 12%. These people could also be willing to lose all their investment if things go wrong. In their case, they would point to a higher risk given the economic attitude they have at the base.

These two different groups of people have different "risk profiles".

It is important that in any purchase you make in your life (even for something "concrete" like a car), you do so knowingly about the financial risk profile you can afford to take.

My personal opinion is that just because something has higher chances of performance, does not mean it is the best choice. In particular, I have invested mainly in the top 5 coins in terms of capitalization, because they are the safest spot right now. However, I always allocate a small part of my portfolio, 10% to be precise, to low cap coins. How do I find the most promising one? Here is what I do.

Step 2 - Identification of new coins or tokens

There are three main ways I usually use to find the "new" coins or tokens:

> 1. Through the posts of the Bitcointalk.org forum, more precisely in the section "Announcements (Altcoins)";
> 2. In the subreddit/r/cryptocurrency;
> 3. In the "Newly Added" sections of Coincheckup and "Recently Added" by Coinmarketcap.

Each of these is a great resource to discover

interesting coins with great return potential over a shorter period of time. As already said, I only put in a maximum 10% of my capital into these underrated projects.

With every investment comes the possibility to get scammed, and in the crypto world, it happens more often that I would like to see. During the last three years of experience, I have developed a series of principles that I follow in order to avoid being scammed. Here is what will make me decide NOT to invest in an asset.

Step 3 - Exclusion of coins and useless tokens/scams

One of the first things I do when I look at new projects is to subject them to very strict criteria to remove "fluff" projects from the list. In particular:

> 1.   I do not buy cryptocurrencies in industries and sectors that I do not understand;
> 2.   I do not buy cryptocurrencies whose teams are inactive in social media communication;
> 3.   I do not buy cryptocurrencies whose

startups/associations/companies are registered in countries where I cannot validate a solid corporate entity;

4.   I do not buy cryptocurrencies if I cannot find the team members (with particular attention to the founder) on LinkedIn and validate that they are real profiles;

5.   I do not buy cryptocurrencies whose teams adopt spamming strategies and do aggressive and non-informative marketing campaigns on social and non-social channels;

6.   If a team is building a brand new technology, I do not buy the cryptocurrency/token unless there is a detailed technical document explaining how it works;

7.   If a cryptocurrency has a pre-ICO with a discount, I tend not to buy it. If I did, it would only be where the discount compared to the public ICO is minimal and the amount purchased is "locked" for a significant period of time (to avoid massive dumps after the public ICO);

8.   I do not buy cryptocurrencies if I do not use them personally as an end user.

To help me with the process, I also use a series of questions that allow me to get more in depth and realize the true fundamental value of an asset. In particular, I really like to ask myself the following questions:

- Would I use this cryptocurrency as an end user?
- Would I pay that price as a user?
- Does this project require the development of a new technology?
- What is the team's experience in this determined direction? Have they already managed a successful company? What was the performance of this company?
- Does the team have the ability to develop this technology? Are engineers and developers recognized in this sector?

Do they have product managers and customer support?
- Is it clear how the project will generate users/customers?
- Why are they using the blockchain? Do they really need it or do they use the term "blockchain" to hype

their project up? What are the pros and cons of using the blockchain in this case and why should the blockchain improve the current alternative on the market? (Keep in mind that currently, blockchain-based systems are slow and expensive).

Pay attention to absolutist statements. Each project has negative aspects and consequences. A real project will be realistic in delineating them, especially the latter.

If I can see that each question has a positive answer, I will then allocate a part of my portfolio. I always invest long-term and I am willing to stay in a coin for at least one year. If for any reason, I do not feel confident enough to put money into a project for at least 52 weeks, then that means that it is probably better to look at another one.

Predicting the next currency that will make the boom is impossible, out there are so many projects based on nothing that still capitalize tens of billions of dollars. In the same way, there are dozens of serious projects that deserve more, but that fails to stand out and gain visibility compared to others. The golden rule is that which applies in every financial market: diversify. By diversifying between several coins, you reduce the risk.

Chapter 13: 20 Golden Lesson to Trade like a Pro

Now that we have gone through the main mistakes a beginner makes, it is time to take a look at what we call "the 20 Golden Lessons of trading".

1.   *"If you are undecided, stay still"*. It is not necessary to invest continuously. If you do not have precise ideas, it is better to do nothing and wait for clearer signs. Often times, the market is full of indecision: keep calm and stack up money for the future.

2.   *"Cut losses and let profits run"*. This is perhaps the best known and most important rule for those investing in the stock market. An indispensable factor for the application of this rule is the identification, immediately after the purchase, of the stop loss. This is how much you are willing to lose on that investment (take into account when determining the average daily excursion of the stock). The cold and systematic application of the stop-loss, even if painful, will preserve you from

huge losses that would make the sale more and more traumatic, freezing capital that could be invested elsewhere.

3. *"Learn from your mistakes"*. Errors are not always negative: if you follow a strategy with a method and if you apply the stop losses, you will not make particularly serious mistakes. Errors are an integral part of stock trading. You need to analyze why you made them and what you can learn from them. In this way, a small loss can become a good investment lesson for the future.

4. *"Take profit and invest them back"*. If one of our titles is on the rise, take profit will be applied as the stock grows. A stock cannot grow indefinitely. When the trend is reversed, selling at the top, we will have had a profit avoiding further descents. If then the title should go up again, it does not matter. It will go better next time. You cannot always sell at the top since you cannot time the market.

5.   *"Buy on the rumor and sell on the news"*. When positive news on a certain title officially comes out, pay attention. It may already be too late to invest in that title since the market could already have priced it in.

6.   *Do not believe in "safe investments"*. If someone tells you that a title will certainly reach a certain price, he either does not understand much of the stock market or is only doing his own interests.

7.   *"Never become emotionally attached to a stock"*. Some investors always follow a limited number of companies that they consider more reliable than others. There are no titles better than others, but only favorable situations and unfavorable situations. Often, instead of admitting an error, one perseveres on it with the consequence of being heavily unbalanced on a stock. This is really bad, especially if you are overcommitted to a stock in which, at that moment, the market does not believe in.

8. *"Always maintain certain liquidity available"*. Cyclically, we find ourselves in situations of several days of generalized decline of the whole stock exchange and often, for lack of liquidity, we cannot grasp excellent buying opportunities. Keep some money aside to jump on big opportunities.

9. *"Choose the right platform"*. One important rule for investing in the stock market is that the platform makes the difference. Carefully selecting safe, honest, and reliable trading platforms is the first step to make money. Those who start investing in the stock market for the first time must be careful to choose platforms that are really simple to use, perhaps with a high-quality educational support. Some platforms also offer add-on tools such as notifications, social trading, and free analysis tools to guide less experienced traders.

10. *"Invest only in what you understand"*. As the "guru" of finance, Warren Buffett said, "Never, never,

invest in something that you do not understand, and above all, that you do not know". The overwhelming majority of investors can achieve their capital growth goals by using the most common financial instruments, which are almost always simple to understand. The complex tools are best left to the great experts in the field.

11. *"Diversify your portfolio"*. When investing, the word to keep in mind is "diversification". Never invest in a single title, because if that sinks, your money will come to the same end. It is always better to have diversified investments to minimize the specific risks of a company, a market, an asset class or a currency. The more you diversify the lower the probability of having drastic falls.

12. *"Understand and evaluate the risk"*. A risk is an intrinsic component of every investment. If it does not exist, there is no return. Whether they are government bonds, stocks, or mutual funds, they all have a risk component, which will obviously be greater if you want to hope for higher returns. So, if

someone tells you that there is an investment without risk, it means that it is better to get advice from someone else.

13. *"Look beyond direct investment"*. As an alternative to direct purchase of shares, it is possible to invest in the stock market indexes through ETFs (listed mutual funds, which replicate the performance of equity and bond indices) or in mutual funds that offer a high diversification even with minimum amounts. This allows you to invest small periodic shares, for example, 100 Euros per month, and may even provide a monthly coupon.

14. *"Do not follow the masses"*. The typical decision of who buys stocks by investing in the stock market is usually strongly influenced by the advice of acquaintances, neighbors, or relatives. So, if everyone around is investing in a particular company, the tendency of a beginner investor is to do the same. But this strategy is bound to fail in the long run and it is not the right approach. There should be no need to say that you

should always avoid having a herd mentality if you do not want to lose hard-earned money on the stock market. The world's biggest investor, Warren Buffett, is right when he says "Be fearful when others are greedy, and be greedy when others are fearful!"

15. *"Do not try to time the market"*. One thing that Warren Buffett does not do is try to time the stock market, even if he has a very strong understanding of the key price levels of the single shares. Most investors, however, do exactly the opposite, which often causes losses of money. So, you should never try to give timing the market a chance. In reality, no one has ever succeeded in doing so successfully and consistently over multiple market cycles.

16. *"Be disciplined"*. Historically, it has often happened that during periods of a high market upswing, we first caused moments of panic. Market volatility has inevitably made investors poorer, even if the market moved in the intended direction. Therefore, it is prudent to have patience and follow a disciplined investment approach as

well as keeping a long-term general picture in mind.

17.   *"Be realistic and do not hope"*. There is nothing wrong with hoping to make the best investment, but you could be in trouble if the financial goals are not based on realistic assumptions. For example, many stocks have generated more than 50 percent of returns during the big uptrend in recent years. However, this does not mean that we can always expect the same kind of return from the stock exchange.

18.   *"Keep your portfolio under control"*. We live in a connected world. Every important event that happens anywhere in the world also has an impact on our money. So we have to constantly monitor our portfolio and make adjustments.

19.   *"Be sure to be on the legal side of things"*. If someone proposes an investment, it must be verified as an "authorized project". In our country, those who offer financial investments must be authorized by law, and this is

an important safeguard for savers. In fact, the authorization is issued only in the presence of the requested requisites. Once authorized, the financial intermediaries are subject to constant supervision. Checking this is not particularly demanding. If you have an internet, you can even directly access the information held by the supervisory authorities. Otherwise, you can contact the authorities themselves using traditional means.

20. *"Be skeptical and do your own research"*. Nobody gives anything for nothing. Be wary of investment proposals that ensure a very high return. At the promise of high returns, there is usually very high risks or, in some cases, even attempts of fraud. Be wary of "Ponzi schemes" which promise profits linked to the subsequent adhesion of other subjects, who often must be convinced by the investor himself to join. These "operations", in fact, cannot guarantee any kind of return, as they are normally supplied exclusively by the continuity of the accessions. In other words, when the new signatures are no longer

sufficient to pay the "interests" to the previous subscribers, the schemes are destined to fail. Be wary of the vague and generic investment proposals, for which the methods for using the money collected are not explained in detail (what kind of securities will be purchased, at what prices, on which markets, with which risk profiles - interest rate , foreign exchange or counterparty, and whether and which hedging instruments will be used to cover such risks).

Chapter 14: Technical Analysis and Fundamental Analysis

Technical Analysis is the study of graphs. Looking at the charts, the analyst is able to understand if that stock (or market) will rise or fall in a short, medium, and long term. The Fundamental Analysis, on the other hand, bases its forecasts on the "fundamental factors" like news, market rumours, company acquisitions, economic crises, political events, wars, etc.

Which is better between Technical Analysis and Fundamental Analysis? Who has never asked this question? The answer is simple. As always in investments, there is no better. It depends on the investor, on his way of operating in the markets, on his degree of risk, etc. In other words, there are those who are better off with one and there are those who are better off with the other.

I personally love the Technical Analysis much more for some reasons. Let me present some of them:

• Timing: Technical Analysis offers better Timing than Fundamental Analysis. Timing is "the right time to get into a position", the ideal time to enter the market. It is, in my opinion, one of the fundamental concepts to succeed in the stock exchange. If you use the right timing, you can afford a very tight Stop Loss, so you can only lose a little. So cut the losses and let the winnings run, the golden rule of the stock exchange. Timing is obviously given by the key levels that are obtained without problems with the study of the graphs, and then through Technical Analysis.

• Flexibility: Technical Analysis is more flexible than Fundamental Analysis since it gives us key levels (for Stop Loss and Goals) in any time-frame.

• Discount: Technical Analysis discounts the Fundamental Analysis, basic postulate of Technical Analysis. The chart already includes all the factors, all the news, all the wars, all the economic conditions, etc. As a result, if the price has risen, the fundamentals will be bullish. If the price has dropped, the fundamentals will be bearish. I can only take care of the chart, thus eliminating many variables.

In addition to this, Fundamental Analysis has the defect that certain news is difficult to find for a common investor, and sometimes, when this news arrives, they are now useless, because someone smarter than us have already used them and bought (or sold) before us.

We close with a sort of "metropolitan legend" of trading, a widespread belief (but wrong) that many still have today. Many investors believe that the Technical Analysis serves to make investments in the short term and that the Fundamental Analysis serves to make long-term investments. This is not true. Both can be used to operate in the short, medium, and long-term.

So many investors will continue to appreciate one and many to appreciate the other. A good idea, sometimes, is to use both, thus combining the advantages of one with the advantages of the other. An application of this concept has been explained regarding refuge currencies and high-yield currencies in Forex.

Can technical and fundamental analysis co-exist?

Although technical and fundamental analyses are considered as opposite poles, many market participants have made a winning combination. For example, some fundamental analysts use the tools of technical analysis to identify the best times to enter the market.

Nevertheless, many technical analysts exploit the economic fundamentals to support technical signals. For example, if a technical pattern on the chart indicates the possibility of selling, we can refer to the fundamental data to obtain a confirmation of this pattern.

A mix of technical and fundamental analysis is not well received by the "extremists" of both schools of thought, but the benefit we can derive from fully understanding the technical and fundamental analyst's mindset is undeniable.

Conclusion

Thank you for making it to the end of this book. I hope it was able to provide you with all the tools you need to achieve your financial goals.

The next step is to get started with what you have learned during the course of this book. Remember to always start with a demo account. Become a profitable trader before putting your money on the table.

I hope that you find these lessons valuable and that you got the information you were looking for. Creating a "trading lifestyle" that works for you will give you an incredible feeling, especially at the beginning, when you make the first gains. I am thrilled for you to start and I cannot wait to see your results coming in.